Adrift

Tales of Ocean Fragility

Edited by

Claudio Campagna

Yvonne Sadovy de Mitcheson

Nicolas Pilcher

Andrew Hurd

Julie Griffin

This book was made possible by the generous contributions of

This book is a product of

This publication has been made possible in part by funding from the Lighthouse Foundation, the European Bureau for Conservation and Development, the Society for the Conservation of Reef Fish Aggregations, the University of Hong Kong, the Marine Research Foundation, the Wildlife Conservation Society, the IUCN Species Programme, the IUCN Global Marine Programme, and the IUCN Species Survival Commission. A full list of contributors can be found on the acknowledgments page.

Published by: IUCN, Gland, Switzerland

Citation: Campagna, C., Sadovy de Mitcheson, Y., Pilcher, N., Hurd, A., Griffin, J., Editors (2008). Adrift, Tales of Ocean Fragility. Gland, Switzerland: IUCN. 136pp.

ISBN: 978-2-8317-1070-9

Cover design by: Eugenia Zavattieri

Cover photo: Turtle by Nicolas Pilcher, Grouper fish by SeaPics

Layout by: Eugenia Zavattieri

Produced by: Marine Conservation Sub-Committee of the IUCN Species Survival Commission, IUCN Species Programme, and IUCN Global Marine Programme

Printed by: SRO-Kundig, S.A., certified vendor of FSC (Forest Stewardship Council) products

Available from: IUCN (International Union for Conservation of Nature)
Publications Services
Rue Mauverney 28
1196 Gland, Switzerland
Tel +41 22 999 0000 /Fax +41 22 999 0020
books@iucn.org
www.iucn.org/publications
A catalogue of IUCN publications is also available.

The text of this book is printed on 170gsm Satimat Green FSC Mix.

A message from the Director General of IUCN

Some might wonder what motivates us at IUCN to take on the daunting task of nature conservation when we live in a world that may seem to be driven blindly forward by the economic bottom line, and in which ceaseless bad news from the media might have us despair at the plight of the planet. It can be easy to forget that in the oceans that surround us, a range of fascinating animals and plants drive not only many economic markets but also many life-sustaining ecological processes. This is our motivation, and in our role as a worldwide union to conserve nature we have a responsibility to share this inspiration with others so that they will join us in our mission.

My hope is that through this compelling book you will take away stories of incredible wildlife spectacles, and think again about our impact on the oceans. Far from being a cry for complete denial, if anything the stories herein call for wise use and appreciation of the diversity of life in the hope that we will never cease to wonder at the oceans that form such a large part of our planet.

I trust you will join me in thanking the people who have taken the time to draw our attention to these living legends.

Julia Marton-Lefèvre
Director General, IUCN, the International Union for Conservation of Nature

A message from the Chair of the Species Survival Commission

Species – the plants and animals that are the building blocks of nature – easily win our admiration but their diversity requires special attention in conservation work. The IUCN Species Survival Commission (SSC) – the largest of IUCN's six volunteer Commissions – turns 60 years old in 2009. It has a proud history of species conservation, including developing the IUCN Red List of Threatened Species™ and tools and guidelines for conservation practitioners. The SSC has over 7,500 members in more than 100 Specialist Groups, some of which focus on conservation issues for a specific group of species, while other groups tackle broader issues such as sustainable use and the impacts of alien invasive species.

In 2005, the SSC formed the Marine Conservation Sub-Committee (MCSC), co-Chaired by Yvonne Sadovy de Mitcheson and Claudio Campagna, to enhance collaboration between IUCN's marine experts within SSC, the Species Programme, the World Commission on Protected Areas, and the Global Marine Programme, and key partners such as the Food and Agriculture Organisation of the United Nations (FAO), and TRAFFIC. This group acts as a crucial focal point for marine issues, providing an advisory role for the Global Marine Species Assessment and taking on specific marine issues not covered by other components of IUCN.

Recognising the lack of awareness and action regarding the current over-exploitation of marine species, the MCSC produced this book to bring priority marine conservation issues into the public eye through a series of compelling stories.

Congratulations to the SSC MCSC for bringing to our attention these wonderful illustrations of conservation in action, and for taking a pragmatic look at the work that remains to be done.

Dr. Holly T. Dublin
Chair, IUCN Species Survival Commission

A message from the Editorial Team

We set out to change the course of history. Ambitious for a book, you may think, but we wanted to bring you something different, something unique. And we are presumptuous enough to hope that you will take away a message that goes more than skin deep, that fills you with the concern and passion that drives us, and leads to significant changes for a better planet.

Our vision was also a bit different. Numerous campaigns have been launched to save the oceans, the whales, turtles, seahorses, and much more. But often the best awareness campaigns have fallen on deaf ears, or were drowned by the sheer volume of varied plights across the globe and under its seas. We wanted you to experience the diversity of life in the oceans, the conservation challenges, and the some of the solutions and success stories.

We chose the 'cocktail party book' approach to put these stories into a more informal setting, so you would be awed and inspired to spread the word. The marvellous tales assembled here are testament not only to the eccentricity of our marine life, but also the diversity of challenges and opportunities we face to conserve these spectacles.

By nature we are optimists and we believe there is a future out there for us to coexist harmoniously with all other creatures. From tales of curious sharks to body warming ducks in the Arctic, through ghost fishing on the high seas and the quintessential underwater gatherings, we hope you come away from this experience feeling as fascinated and concerned about our oceans as we do.

Claudio Campagna, Yvonne Sadovy de Mitcheson,
Nicolas Pilcher, Andrew Hurd, and Julie Griffin
The Editorial Team
SSC Marine Conservation Sub-Committee

Table of Contents

Hanging at Sea

A thousand mature elephant seals are gathering along a wind-swept beach in coastal Patagonia. It's breeding season. Huge males – some weighing nearly three tonnes – will crash together like small trucks in fights for breeding rights. Females, much smaller, will dodge the combatants, submit to the winners, and produce pups from the previous year's mating. It's a drama that must be played out on land before the seals can go home to the ocean's depths.

Hanging at Sea

The southern elephant seals that haul up on that Patagonian beach are testament to the toughness of the species. But, even this vast expanse of ocean and coastline called the Patagonian Shelf, home to some of the most awe-inspiring wildlife spectacles on earth, rivaling terrestrial counterparts like the wildebeest migration in eastern Africa, is no longer immune to the economics that drive commercial fisheries around the world.

Starting in 1995, a few males and females, youngsters and adults, arrived on the beach with neck wounds in various stages: fresh and bleeding or infected gashes, or scars. They had been caught – or more accurately, hanged – in fishing line. Though certainly in a class by themselves when it comes to grace under water, elephant seals are unfortunately not alone when it comes to encounters with fishing gear. Fish large and small, sea birds, marine mammals, turtles and just about any other ocean creature are found entwined in fishing gear in every ocean, every day.

From 1995 to 2005, at least 35 seals were seen either with fresh wounds or scars around their necks. The culprit was fishing line, identified when it was cut away from the necks of 18 seals. In every case, the strangling line had a circle tied with a human-made knot.

Hanging at Sea

The origin of the line remained a mystery until two entangled juveniles carried "jigs" – coloured lures armed with a crown of hooks – on the line wrapped around their necks. Jigs are gear used to catch squid – and commercial fishing for Argentine squid had begun at just about the same time as the entangled seals started to be recorded by researchers. Mystery solved. The unlucky seals had come into close contact with the fishing vessels or their debris.

Of the 18 elephant seals "rescued" by cutting away the fishing line, only five were sighted in subsequent years. Much can happen in the seas, but it is reasonable to assume that at least some of these animals ultimately died from their wounds.

Fortunately, improvements in fishing gear and methods are starting to cut down on the number of animals and birds being entangled, but much more work needs to be done to ensure our human species can continue to enjoy the fruits of the sea without completely destroying marine life in the process.

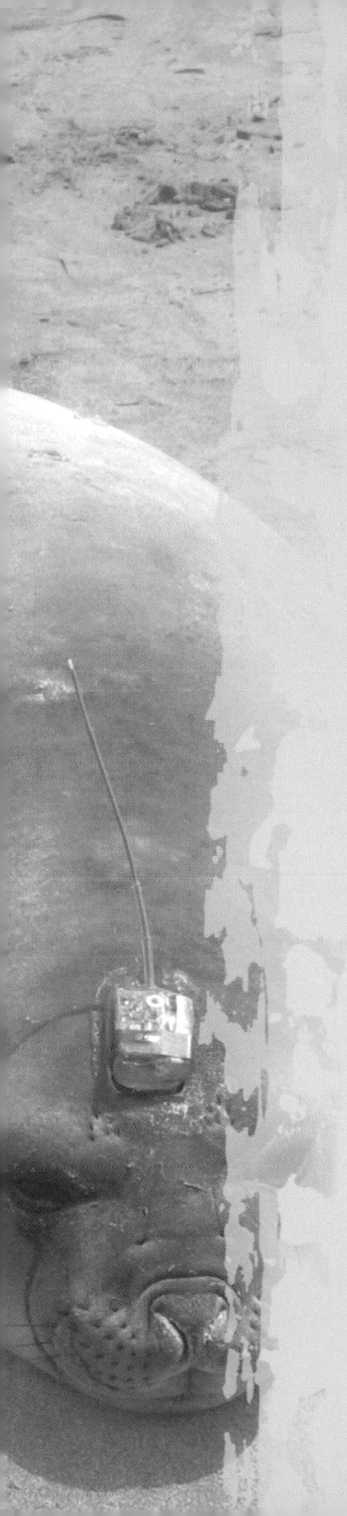

Olympic Class Athletes

There are two species of elephant seals, one in each hemisphere. When they come on land to reproduce, they look like clumsy aliens, but in the ocean these animals are Olympic-class athletes. These marine mammals breathe air in their spare time, so to speak – when at sea they remain in the silent, dark, deep at least 90 percent of the time; they can make dives down to depths of several kilometres to feed.

And in their most Herculean feat, elephant seals endure: hunted and turned into oil for lamps and other human uses, they were at the brink of extinction less than a century ago. They fought back. Today, most colonies are stable or recovering from population declines that lasted decades. The Patagonian colony of Southern Elephant Seals (Mirounga leonina), *is the only one that has expanded, at least during the 1990s.*

Drugs, Money, Madness – and a Mollusc

You don't need money to score methamphetamine in South Africa, you need abalone.

The illegal trade in wildlife – as products, pets, or food – is a US$20 billion a year global business topped only by trafficking in guns and drugs. In some cases the symbiosis of the wildlife, weapon, and drug trades creates a perfect storm of shadow enterprise and organised crime.

MCM * VEROORSAAK POACHING
MINISTER ONS SIEN JOU IN HOF
MINISTER BEDANK ONS SOEK 'N LEIER NIE 'N VERLEIER
DAY 1

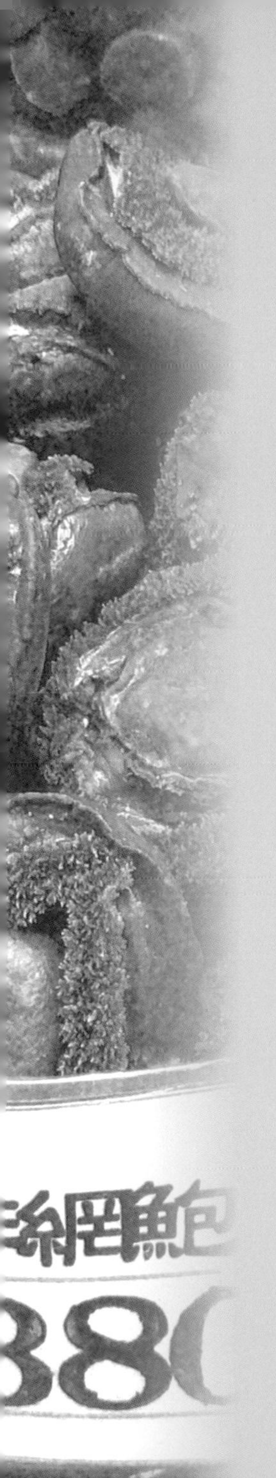

That perfect storm landed in Cape Town several years ago. The South African government has retaliated with high-speed naval vessels, divers and other special enforcement troops, helicopters and surveillance aircrafts, a special court, and dogs that are trained to sniff out the contraband – the humble abalone. The plate-sized ocean mollusc is so prized as a food delicacy it's worth about US$500 to US$800 per kilogram on Asian black markets. A five-star abalone meal in China sets diners back several thousand dollars per person.

Chinese gangs moved into South Africa to rule the multi-million dollar illegal abalone trade and a specialised drug market in which abalone is the currency. The mollusc is swapped for ingredients to make methamphetamine, the highly addictive crystallized amphetamine known locally as tik*. Abalone poachers who forgo the tik makings can opt for good old-fashioned cash. The overall trade attracts South Africans of all ages, from young school dropouts to adults who give up legal jobs for a piece of the abalone action.*

Drugs, Money, Madness – and a Mollusc

Those who get their tik through the trade are helping Cape Town set new, dubious records: drug-related crimes have risen 200 percent in the past two years, and according to hospital records there is a significant rise in emergency cases of psychosis brought on by chronic use of tik.

The planes, dogs and divers have done little to calm the storm nor has international assistance in the form of trade monitoring and regulation. In January of 2008, the legal abalone fishery closed, but this was in the face of unrelenting poaching and illegal trade. The problem is by no means limited to South Africa: Californian and Canadian commercial fisheries for other types of abalone have been closed for many years due to pressure on stocks from illegal harvest and trade.

The toll on abalone worldwide remains intolerable. There may be no other marine animal that suffers such high levels of poaching and illegal trade, and South African abalone continue to bear the brunt of the onslaught.

If the trade is not brought under control, it is likely to cause the commercial extinction of South African abalone and the loss of an important traditional livelihood for many South African coastal communities.

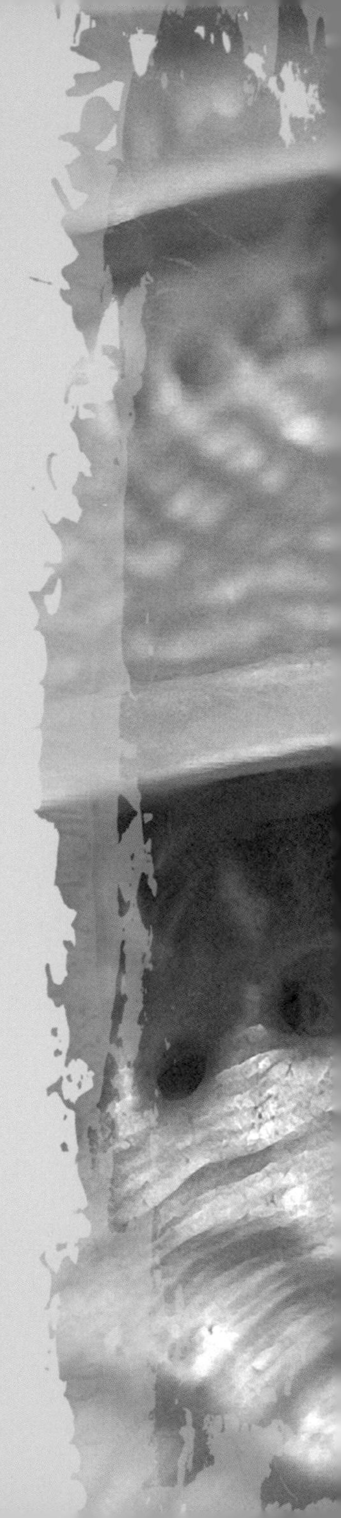

Stop Selling These Sea Snails

Abalones, or perlemoen in South Africa, are marine snails best known to many people for the iridescent mother-of-pearl coating on the inside of the low open spiral shell of many species. There are over 100 species worldwide.

The endemic South African Abalone (Haliotis midae), *grows up to 20 cm and can live more than 30 years. It grows slowly, taking about seven years to become mature and reproduce which means that it cannot withstand heavy fishing pressure.*

In 2007, South Africa listed the species in Appendix III of the Convention on International Trade in Endangered Species (CITES) providing international assistance to regulate trade, and fishing in South Africa was banned in early 2008 to stem rapid local declines. The ban itself is controversial because illegal harvest, not legal take, is the main cause of a decline in stocks and poaching has continued despite the fishing ban. Abalone farming and hatchery production provide hope for the future although wild hunting will likely continue as long as it is the cheaper and easier option.

Fatal Attraction for Fish

It doesn't get much worse than this: when you have sex, you're likely to die.

Imagine you're a male Nassau Grouper – a big, charismatic fish of the Caribbean reef. Normally, you're a loner but today you find yourself heeding a feeling that you get just a few times a year – you've simply got to find other groupers and fertilize some eggs. As dusk approaches and a full moon starts to rise, you eagerly cut through the warm water to the place where you know you'll find the action.

Fatal Attraction for Fish

Sure enough, as you cruise near the reef edge you can see the party's already started. Thousands of groupers are here. Packs of males are chasing single, egg-laden females up to the water's surface; white clouds of eggs and sperm are going off around you and above you like fireworks in the darkening sea.

The party on the reef is what's known as a spawning aggregation, in other words, a reproduction extravaganza. These gatherings can number tens of thousands of individuals, some travelling hundreds of kilometres to participate each year. But uninvited others also come to crash the party.

Fish come to feast on the billions of released eggs and sharks prey on the distracted adults. But this natural predation pales in comparison to the toll taken by commercial fishing. Fishermen have learned to target the traditional spawning grounds at just the right time - in just a few hours they can, and sometimes do, catch nearly all the fish drawn to the place. The aggregations, which typically form at the same times and places each year, are being rapidly discovered and depleted.

Fatal Attraction for Fish

Fatal Attraction for Fish

This fatal attraction between fish and humans is played out not only with groupers but also with many other commercially valuable reef fishes.

The case of the Nassau Grouper is a prime example of the problem. Historically enormous and numbering tens of thousands of adult fish, many aggregations of this species today draw only a few hundred individuals and some no longer form at all. The brief gatherings every few months add up to fewer than five hours a year for groupers to reproduce. Without these opportunities, populations collapse.

The Nassau Grouper was classified as endangered on the IUCN Red List in 1996 and again in 2003, yet their numbers continue to decline. Protective legislation was introduced in several countries to reduce pressure on the species, but enforcement continues to be a major problem. Most aggregations still receive little or no protection – and life continues to be cut off at its sexiest moment.

Glory Days Over for Groupers

The Nassau Grouper (Epinephelus striatus) *was once the most commonly landed fish on Caribbean islands. Although its fishery is now much reduced, its flesh is still highly regarded and fetches top prices in restaurants.*

Fewer than 100 spawning aggregation sites are thought to occur for this species and more than half of those we know of have disappeared because of fishing. Once the aggregations go, the populations they service dwindle and collapse.

A shadow of its once glorious self, the Nassau Grouper has been demoted from major reef fishery species to a fish threatened with extinction due to uncontrolled fishing. And this happened over just a few short decades. Although country-level management has few success stories to boast, the good news is that regional measures in development hold promise for restoring not only the sex life of groupers but their numbers as well.

Waste and Horror at Sea

In the time it takes you to read this sentence, about four tonnes of fish are being hauled out of oceans – and thrown away. Over the next 24 hours, the total discards will rise to nearly 19,000 tonnes (that's the equivalent in weight of 4,000 adult elephants, dumped overboard, each day). And in a year, more than 7 million tonnes of fish and other ocean animals will have been caught and dumped back, dead, into our oceans and seas (yes, that's correct, nearly equivalent in weight to 1.5 million elephants each year).

Waste and Horror at Sea

It all adds up to this: roughly 25 percent of the total annual catch of the world's fisheries is thrown out on the spot as garbage.

These billions of animals, including tonnes of edible fish, are bycatch: creatures caught by humans who were trying to catch something else. Bycatch includes fish that are too small, species that are not a commercial fleet's target animal, and other animals simply caught in nets or on lines. Nearly all discards die – tossed dead or dying back into the sea for hungry gulls and other opportunists who, that day at least, managed to avoid becoming collateral damage.

Certain types of commercial fishing methods are responsible for most bycatch. Trawling or dredging, often used for shellfish, can destroy sea floor habitat and catch everything in a ship's path, intended or otherwise. Long-line fishing involves a system of lines and hooks that can stretch up to 50 miles across the ocean. Drift nets 10 to 15 metres deep are unanchored and often suspended across 90 kilometres of sea to scoop up anything bigger than their mesh size. These and other similar kinds of gear do not discriminate: fishes of all sizes and species, dolphins, porpoises, sea turtles, gulls and other sea birds, seals, crabs – even whales and great white sharks – become fatally entangled in nets or caught on hooks.

The numbers are staggering:

For each kilogram of shrimp caught in a trawl net, an average of 10 kilograms of other marine life is caught and thrown overboard as bycatch.

As many as 300,000 whales, dolphins and porpoises die as bycatch each year.

Nearly 100,000 albatrosses are among the sea birds that drown each year, when they dive for bait on commercial fishing lines, are hooked and pulled under.

A partial list of bycatch from one drift net fleet hunting squid in the Mediterranean in 1989: 58 blue sharks, 914 dolphins, 52 fur seals, 35 puffins and 22 marine turtles.

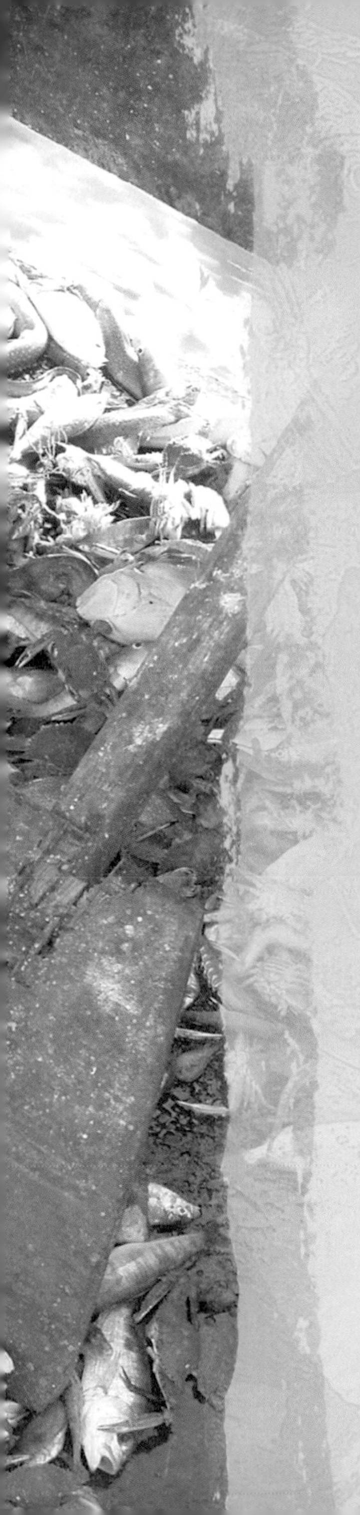

Drift nets are now banned globally in international waters but still used illegally in some places, and the proof is that they are regularly abandoned or lost by fishing fleets. Miles and miles of these "ghost nets" are floating unchecked through the oceans right now, effortlessly ensnaring and carrying a ghastly catch of dead or dying marine life. It only stops when the weight of the snagged animals drags the netting to the ocean floor.

The news isn't all bad. Progress is being made in some fisheries and for some species. Devices have been tailored, such as 'turtle-excluders' and specially designed hooks and fishing methods to reduce such unintended catches of turtles, seabirds and dolphins. Bycatch of albatross, that magnificent bird that glides so effortlessly over our oceans, can be significantly reduced if fishing vessels employ streamer lines to dissuade the birds from coming too close to the boat where they are most vulnerable, or set their hooks to reduce opportunities for interaction between hook and bird.

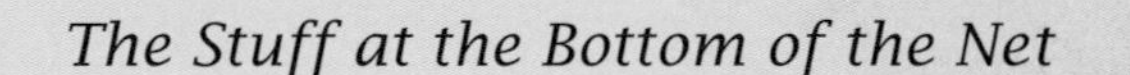

The Stuff at the Bottom of the Net

When big and charismatic animals are incidentally caught and killed by commercial fishing fleets, we easily grasp the tragedy of bycatch. But public attention and even scientists have largely ignored the impact on hundreds of small fish species about which we know virtually nothing – the "stuff" at the bottom of the nets.

Waste and Horror at Sea

Thousands of tonnes of these fish become anonymous victims every day. Any research to date has focused on numbers and weights, neither of which indicates the ecological impact of such devastation. These small creatures play vital roles in ocean communities as prey, competitors, or predators, and are key determinants of how other populations fare.

Common sense, supported by scientific understanding, requires that we keep these building blocks in place if ecosystems are to function effectively. But we risk never even knowing what that stuff is, or was, at the bottom of the net – such waste is unethical, and such ignorance is dangerous for the future of our oceans and the people that depend on them.

Dried Shark

Man Eats Shark, World Yawns

That's essentially the situation for shark species that are being fished and finned into oblivion before most of the world knows they even exist. To paraphrase the observation made by the calculating mayor in Jaws, "You say great white shark and you have a panic. You say porbeagle and people say, huh?"

Even diehard fans of sharks might not be familiar with the Porbeagle Shark – a big, fast relative of the Great White that has the misfortune of being a luxury menu item rather than a Hollywood icon.

Fisheries target the species since Porbeagle is among the most prized of all shark meat. Even if you have never ordered a Porbeagle steak, you may have unwittingly sampled the fish: restaurants have been known to cut shark meat into appropriately-shaped chunks and serve them as scallops, or to offer it as 'swordfish'.

A major threat to Porbeagles and many other sharks today is the soup pot, as demand for shark fin soup remains high, especially in Asian countries. "Shark finning" yields the key ingredient: sharks are caught, their fins cut off, and then, often still alive, they are tossed back into the ocean to drown or be eaten by other predators. It's better business to fill a hold with valuable fins rather than keep the whole fish.

All sharks are particularly susceptible to overexploitation because they generally grow slowly, mature late, produce few young and live for a long time. For Porbeagles, intense fisheries and high commercial value add to the pressures, and make the species Vulnerable according to the IUCN Red List. Closing or suspending fisheries doesn't necessarily work – the hunt just moves to other oceans and populations, chasing them down one by one. The targeted hunt and shark finning along the way is a scenario more fitting for a horror film than the commonly held nightmare of shark attacks on humans.

And yet there have only been 2,200 shark attacks in recorded history (from 1580 to 2007), while experts estimate that each year millions of sharks are gruesomely and wastefully finned and tossed overboard.

Man Eats Shark, World Yawns

12"-14"洗淨大勾翅
Dried Shark Fin
每磅$258.00/LB.

Man Eats Shark, World Yawns

Sharks have existed for more than 400 million years. They were cruising the oceans nearly 200 million years before dinosaurs roamed the earth. Nowadays, in under 10 years commercial fishing can wipe out a shark population that will then need nearly a century to rebound – if it can come back at all.

Sharks have long needed better public relations. The legendary Great White is helping, as our fearful fascination with these animals slowly turns into appreciation and the desire to protect. But despite the big business – legal and illegal – in shark fins, meat, and other parts, restrictions on international trade are only in place for a handful of relatively well-known species, such as the Basking and Whale Sharks, and Great Whites.

Take Porbeagle management as an example: all data show severe declines in North Atlantic populations, but despite strong scientific recommendations, no international regulations are in place for the species. Only a handful of fishing boats rely primarily on Porbeagle for their income, and yet they have managed to stop policy-makers in Canada and Europe from protecting this shark.

In fact, Europe's failure to put in place regional protective measures in turn undermined its own recommendation to give the species international protection by listing it on the Convention on International Trade in Endangered Species' (CITES) Appendix II in 2007. What will it take for scientists to be heard over the lure of a few dollars? Man eats shark, world yawns.

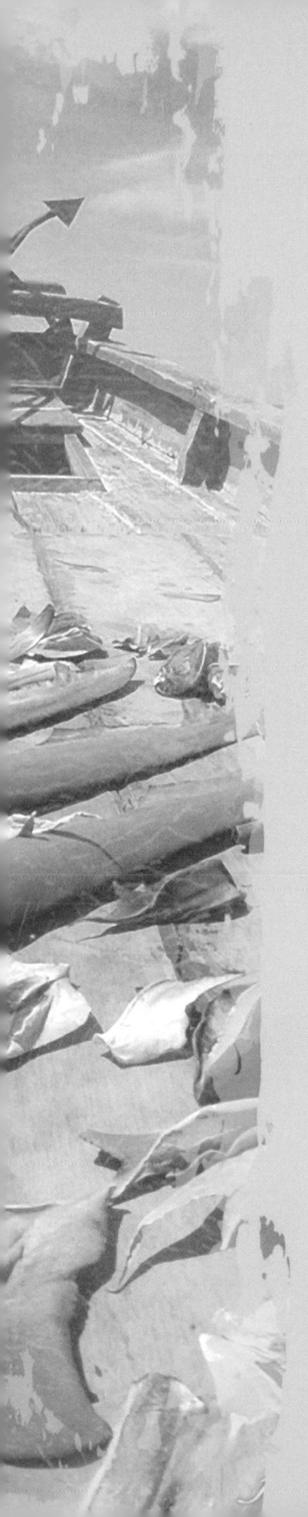

Shark Fin Soup

Shark fin soup is a popular Cantonese delicacy typically served at special occasions such as weddings and banquets at Chinese restaurants around the world. The soup symbolizes wealth and prestige and the fins are valued because they are rare and expensive. Shark fins are appreciated for their texture and ability to absorb flavours but have little flavour of their own.

Chasing the Shark

Porbeagle (Lamna nasus) *population crashes from overfishing in the Atlantic Ocean began in the 1960s. When shark numbers dropped in the Northeast Atlantic, the fisheries simply moved to the Northwest Atlantic and a similar crash followed.*

Sharks are slow to rebuild their populations. Today, in the Northwest Atlantic improved management has stopped the decline, but recovery could take more than 70 years. The Northeast Atlantic population that had first collapsed 40 years ago remains the most depleted Porbeagle population.

Ducks Huddle Up to Create Secret Arctic Hideaway

Ducks Huddle Up to Create Secret Arctic Hideaway

It was a freak signal from a long-silent satellite transmitter that ultimately gave them away.

For decades, Spectacled Eider Ducks have been studied on their breeding grounds. But even with new tracking technology that seemingly removes nature's every fig leaf, no one could figure out where the birds spent the winter. Then, in 1995, that freak satellite signal led biologists to fly over the Bering Sea and finally discover a wildlife spectacle hiding in plain sight: nearly 150,000 Spectacled Eider Ducks, half the worldwide population, hunkered down in a few open pools among the sea ice. Paddling among the ice floes, they could dive down to 70 metres to feed on clams.

Ducks Huddle Up to Create Secret Arctic Hideaway

Historical breeding range
Current breeding range
Molting areas (July - October)

Ducks Huddle Up to Create Secret Arctic Hideaway

Even more amazing was how those open water leads were being kept open: the ducks appeared to be doing it themselves, with the heat of their densely packed bodies keeping the water from freezing beneath them. Alone, a duck wouldn't stand a chance at melting through thick winter ice, but when tens of thousands of warm bodies huddle together, they create an ice hole that provides a winter home and a gateway to feeding grounds.

Now that the mystery of the wintering eiders has been solved, the question is how long this wildlife spectacle will continue. It is a complex set of processes that ultimately leads to more than 150,000 ducks huddled together for months hunting clams.

The Bering Sea's food web depends on the extent, timing, and thickness of seasonal sea ice. In turn, that ice 'life' influences the formation and location of a tongue of cold water that extends from the Arctic down into the Bering Sea. This coldwater tongue is a temperature barrier that affects the entire food web and ultimately the Bering Sea ecosystem that supports not only the eiders but also walruses and other marine mammals. Increases in water temperatures of even one or two degrees can change the entire picture, and that is happening now.

Warming waters are leading to a decline in ocean floor species like clams that are the eiders' primary winter food. Changes in seasonal ice influence distribution of fish; and commercially valuable species like flatfish are beginning to move northward with the warmer waters. Following them could be large-scale bottom trawling fleets that would further jeopardize this previously unexploited habitat.

Just two years ago field scientists acknowledged that warming temperatures and loss of sea ice had caused the northern Bering Sea to shift from an arctic to a sub-arctic system. And it is widely believed that winter sea ice will continue to deteriorate over coming decades, leaving a dramatically changed seascape. In all likelihood, it is one that would not support a large population of ducks that huddle up to melt their own fishing holes to survive the winter.

Unleaded Ducks Only, Please

The Spectacled Eider Duck (Somateria fischeri) *breeds on the remote Arctic tundra of Alaska and northeast Siberia. After mating, then moulting in shallow coastal areas, the birds migrate offshore to remote wintering areas on the sea ice, where they gather to seek the rich food supplies of the Bering Sea. Seemingly far from human civilization, this species faces local risks such as the toxicity of lead shot left behind by hunters and mistakenly ingested by the ducks. Although only steel shot is permitted now, lead poison probably contributed to the decline in the spectacled eider population.*

Searching for Weapons of Mass Destruction? Visit the Nearest Coral Reef

Too many reports about the current state of coral reefs make for eerie reading – they're like news dispatches from some underwater Sin City.

Reefs are being blown up, cut up, poisoned, smothered, beaten, or ripped apart by hand. Coral that survives the violence is left behind to die. Resident fish are killed outright or drugged up and kidnapped live into an international market. Stories from the reef front literally involve murder, mayhem, child slave labour, bribery, and government corruption.

Searching for Weapons of Mass Destruction?

It all makes climate change and coral bleaching or damage from tsunamis seem like kinder, gentler dangers. While this could not be farther from the truth, intentional coral reef destruction is a lot farther from our minds than it should be.

Tropical coral reefs are the rain forests of the sea, the most species-rich ocean ecosystems that host up to two million kinds of marine life. These ancient massive structures also support multi-billion dollar economies – global tourism markets depend on coral reefs for their recreational value, fisheries depend on coral reefs for their productivity and safe access, coastal cities and developments depend on reefs for their shoreline protection.

But increasingly reefs are asked to give up more than they can afford. Humans are deliberately assaulting reefs worldwide in ways that are often illegal and unsustainable. In the Indian Ocean and throughout the Pacific, corals are mined for building and road construction materials, medical and health products (such as calcium supplements), jewelry and knickknacks. Live coral is harvested and traded for professional and amateur aquariums.

Searching for Weapons of Mass Destruction?

Searching for Weapons of Mass Destruction?

Searching for Weapons of Mass Destruction?

Explosions also figure into fishing practices that incidentally kill off more coral. Dynamite or cheaper, easily homemade, Molotov cocktail-type explosives are used in blast fishing – they detonate in the water and fishermen collect the dead or stunned reef fish that float to the surface. The bombing shatters individual coral structures and leaves huge craters in a reef. Full recovery is rare: reef fragments on the ocean floor rock back and forth with water movements and are often too unstable to provide a foundation for new corals to build upon.

In Southeast Asia, local young men and teenagers are hired to dive deep into the ocean in huge groups, to pound apart coral with rocks or pieces of pipe and drive the fish they flush out into big, weighted nets nearby. Known as muro ami*, this practice, now banned but still practiced, came under international fire in 1986 when the graves of nearly 100 muro ami divers – who had been caught in their own nets like bycatch and drowned – were discovered in the Philippines. Many were children 10 years old or younger.*

Direct destruction of reefs exacerbates the challenges tropical corals already face from climate change and the slight increases in ocean temperatures that, alone, can be catastrophic for corals. Scientists estimate that 18 percent of the world's coral reefs were severely degraded in 1998 alone – up to that point, the warmest year on record. From the combined impacts of climate change and local threats, the IUCN Red List deems that 30 percent of the world's reef building coral species are now threatened with extinction.

Who's Exporting, Who's Importing?

According to data from the Convention on International Trade in Endangered Species in 1997, the top exporters of coral – dead or alive – were Indonesia, Fiji and the Solomon Islands. Hong Kong, Korea, and Taiwan were the top exporters of "worked" precious coral for items like jewelry. The data showed the United States as the top importer of live coral and "reef rock," bringing in more than 80 percent of the live coral trade, and importing more than 50 percent of the reef fish sold live to aquariums worldwide. Other major importers included Germany, France, the Netherlands, Great Britain, Japan and Canada. The good news is that trading limits are being slowly introduced and many public aquaria use artificial coral in their displays, but much more needs to be done to keep coral ecosystems healthy.

Reaching New Heights or Sinking to New Depths?

Reaching New Heights or Sinking to New Depths?

When one of the ocean's fastest, fiercest fish is more often found in a can than darting through the waves, something is not right.

While cans of tuna fish in supermarkets and plates of tuna sashimi in restaurants are seemingly infinite in number, the fish themselves are not – even though on paper, tuna would seem to be made to last.

Like some kind of underwater James Bond, these huge, svelte top predators possess a number of tricks up their sleeve to enable them to escape larger predators and ensnare prey. Certain species swim at speeds of up to 70 km per hour by retracting their dorsal and pectoral fins into special slots to minimize drag. And when the water gets cold, the cold get going – by using a special blood supply system to certain muscles that allow tunas to maintain a temperature higher than that of the ambient water ('warmbloodedness').

Tunas are considered highly fecund – in other words, they're good at making little tunas and replenishing their numbers. Unfortunately, high fecundity does not necessarily mean that the species can withstand uncontrolled fishing pressure; declines in populations of many tuna species are teaching us this hard lesson.

One look at the numbers and it's not surprising that most tuna species are in trouble. Most modern fishing for canned tuna started in the early 1940s. Annual catches have increased from fewer than half of a million tonnes during the 1950s to today's estimated yearly take of 6 million tonnes.

The global catch for the principal market species of tuna – Skipjack, Yellowfin, Bigeye, Albacore, and Bluefin – grew ten-fold in just five decades. Now, the overall global market is estimated at $US 5 billion, and a fine fish fetches an incredible amount of money: a single Bluefin Tuna recently sold for a record $174,000 at auction in Japan's sashimi market.

Reaching New Heights or Sinking to New Depths?

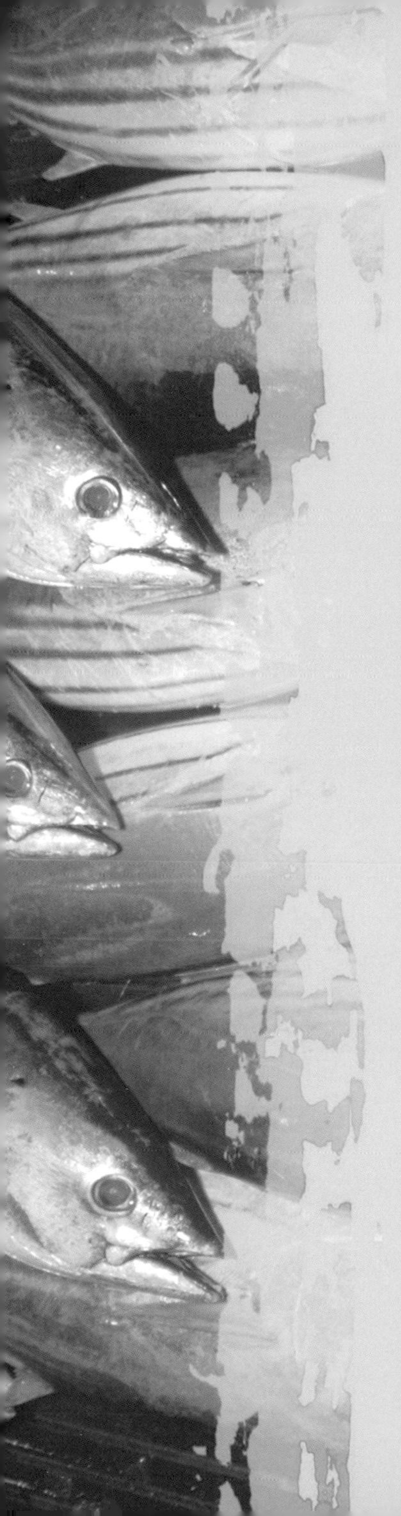

The Mediterranean is one spawning ground for Atlantic Bluefins that sophisticated palates prefer, and fishing tactics there have gone airborne: to find the largest schools of fish churning the waters, fishing operations use spotter planes to scan the seas and radio sightings to ships lurking nearby. These extreme fishing measures are banned in the region to protect the threatened Bluefin, but the planes continue to be used illegally. This may partially explain why the annual quota for Bluefin Tuna was reached in the Mediterranean after only six months of fishing – our ability to catch these fish has reached new heights.

Most tuna regional fisheries management organizations have adopted measures to manage capacity levels and bycatch concerns. But these measures are no match for the growing demand for fish and continued growth of fishing fleets that prevent rebuilding needed for some tuna stocks. James Bond has always stayed ahead of the game; let's hope tunas can do the same.

Reaching New Heights or Sinking to New Depths?

Farming could be a future answer to preventing a total collapse of tuna populations, but not yet. Tunas cannot be hatchery-produced so the only 'farming' now possible is a controversial process involving the fattening of wild-caught lean Bluefin Tunas to increase fat content. The process yields tuna more attractive to the Japanese sushi market but its commercial success puts further pressure on already stretched stocks of wild Bluefins.

The problem is that catching the smaller tunas allows fisheries to take many more fish from the wild and still remain under assigned quotas, because quotas are based on weight of fish and not on numbers. Moreover, a substantial proportion of tuna caught for ranching simply goes unreported.

The world's appetite for tuna seems to have no limits but tuna populations most decidedly do. Wholesale restructuring of regional fisheries management organizations is needed to eliminate the current tendency for scientific advice to be ignored. Another solution may lie with conscientious consumers, who can demand that their tuna only come from sustainable fisheries.

More Than an Entrée

More Than an Entrée

Think about this the next time you sit down to a lobster feast: in happier times, your dinner was the one feasting, not the one being feasted upon.

Although now vanquished, your lobster was once a valiant warrior who wielded his antennae like swords to fight off enemies while undertaking a wildlife migration as spectacular as any in the air or on land.

If autumn leaves are falling, somewhere lobsters are marching. The annual migration of the Caribbean Spiny Lobster to escape stormy shallow waters and move to calmer, deeper feeding grounds translates into tens of thousands of lobsters on the move, concurrently ...and in single file.

Like race car drivers, spiny lobsters draft each other to reduce drag as they walk in head-to-tail queues. The longest known lobster migrations are undertaken by Ornate Rock Lobsters: adolescents march hundreds of kilometres from nursery regions to areas where they mature, mate and spawn.

When caught in the open by predators during these migrations, Caribbean Spiny Lobsters react with classical military tactics and precision. They position themselves to form defensive rosette shapes, and as a unit target the enemy with their spiny antennae 'swords'.

While travelling the seas, spiny lobsters use both a magnetic directional 'inner compass' and a geographic position sense – essentially, they have their own built-in Global Positioning System. Studies have shown that Caribbean spiny lobsters are able to determine their own location on Earth even when moved to an unfamiliar location – an ability known as true navigation.

Obviously there is more to lobsters than meets the grill. But their power is in their numbers, and as lobster populations decline from over-fishing the ability of the migrating troops to defend themselves is diminished, too.

In fact, there is growing concern about just how long these animals will be on anybody's menu.

Groupers, sharks, octopus and a host of other marine species like lobster, too. And, off the menu, lobsters are themselves big eaters. They play an important role as keystone predators keeping urchins in check, preventing the phenomenon of urchin barrens: places where the tireless grazing of urchins has left an area devoid of kelp and other life. Similarly, spiny lobsters can prevent explosions in mussel populations that can leave an area carpeted in mussels and inhospitable to any other species.

Most of the world's spiny lobster species are in decline, due to an unprecedented expansion of lobster fisheries during the last few decades. If we want to keep enjoying lobster dinners and maintain healthy ocean ecosystems, it is in our interest to support our lobster troops in their struggle for survival.

Successful Management Role Models

Wild fisheries are the only option to bring spiny lobsters to dinner plates since much about their biology makes them poor candidates for farming.

The good news is that wild fisheries can be successfully managed, as is happening now with the Western Australian Rock Lobster (Panulirus cygnus) *and Red Rock Lobster of Mexico* (Panulirus interuptus). *The management plans have strong local support and realistic annual catch limits, protection measures for breeding stock, and a foundation of sound science to monitor population health and tackle emerging problems.*

Unfortunately the successes in Australia and Mexico remain unmatched. Elsewhere in the world, fishery regulations are often inadequate or not enforced. As spiny lobster populations continue to decline because of over-fishing and loss of coastal nursery habitats, temperate and tropical ocean ecosystems will feel the loss.

The Last Frontier, Destined for Devastation?

In the deep ocean, pillaging is outpacing exploration.

As much as 90 percent of the ocean remains a last frontier that has yet to be seen by humans. But it is increasingly apparent that while we haven't looked, we most certainly have touched.

The Last Frontier, Destined for Devastation?

In search of new target species and aided by new technologies, commercial fishing has been steadily moving into deeper and deeper waters for marketable fish and invertebrates like shrimp and spider crabs. Most devastating among the fleets are the ocean bottom trawlers that act like wrecking balls on deep seafloor habitats scientists know little or nothing about.

Deep-sea corals are the poster animals for the situation. Scientists are only beginning to confirm the importance of these often-ancient corals and the communities they miraculously create in the cold, dark ocean depths. Only a fraction of these reef-like structures, gardens, and forests have been discovered and mapped. But despite all that is unknown about deep-sea corals, one thing is certain: they're in trouble.

It's a bad sign when much of the information about deep-sea corals comes from examining their dead and broken remains, dragged up from the depths along with other detritus and bycatch in trawler nets. It's depressing when deep-sea corals that take centuries to grow can be pulverized to rubble in one day – and it's tragic that whatever biological, chemical, or evolutionary secrets they could have shared are most likely gone forever. Around the world, scientists, some fisheries, and others are now pushing back with calls for protection of worlds we don't yet know.

The devastation to coral has been a by-product of the hunt for marketable deep-sea fish species with tantalizing names such as orange roughy and oreo dory. These and many others, along with the corals, sponges and other deep-sea life have conquered the dark and the cold by biologically taking life nice and slow. They all live a long time and are slow to reproduce. The very ways that these deep sea animals have endured for hundreds of thousands of years make them exceptionally susceptible to intense harvesting: the chances are about nil that a 50,000-year-old stand of coral can come back to its former glory after being bull-dozed by a bottom trawl when that coral grows about a half-inch each year.

Science and commercial interests alike are being led deeper into the ocean with advances in both exploration and exploitation technologies. The prizes include not only living things but oil and precious metals – perhaps even more than found on land – as well as knowledge and potentially life-changing discoveries. The race is on to the bottom of the sea, and how our last frontier fares will depend on our ability to curb the devastation.

Nets of Destruction

The number one threat to deep-sea corals is bottom-contact fishing, including trawling, bottom long lining and the deployment of traps for fish and shellfish. The targets are shellfish and fish, including very valuable deep-water fishes, some of which live more than 100 years. One of the main techniques used to snag this bounty is bottom trawling: boats release huge trawl nets that sink down to the ocean bottom, where they are dragged through the water. The gear at the bottom includes nets that can be 12 metres high and stretch 60 metres across; the mouths of the nets are held open by huge doors that can weigh several tonnes. As this heavy trawl is dragged across the ocean floor on 200 kilogram wheels, anything in its path is scooped up into the netting or crushed under the weight of the gear.

How They Live(d)

Of the 5,080 known corals, about two-thirds are deep-sea corals. They live in silent, dark, cold waters and feed on zooplankton or particles of food sinking down from the ocean surface. Growing extremely slowly – on average maybe an inch a year – they may be among the oldest living organisms on Earth.

Coral reef structures just explored off the southern Atlantic coast of the United States are believed to be tens of thousands of years old, stretching across the ocean for kilometres and rising up to 91 metres in places from the ocean floor. With their ancient built-in memory, the oldest deep-sea corals can contain important clues about ocean conditions from centuries ago, and may help us map our future for centuries to come.

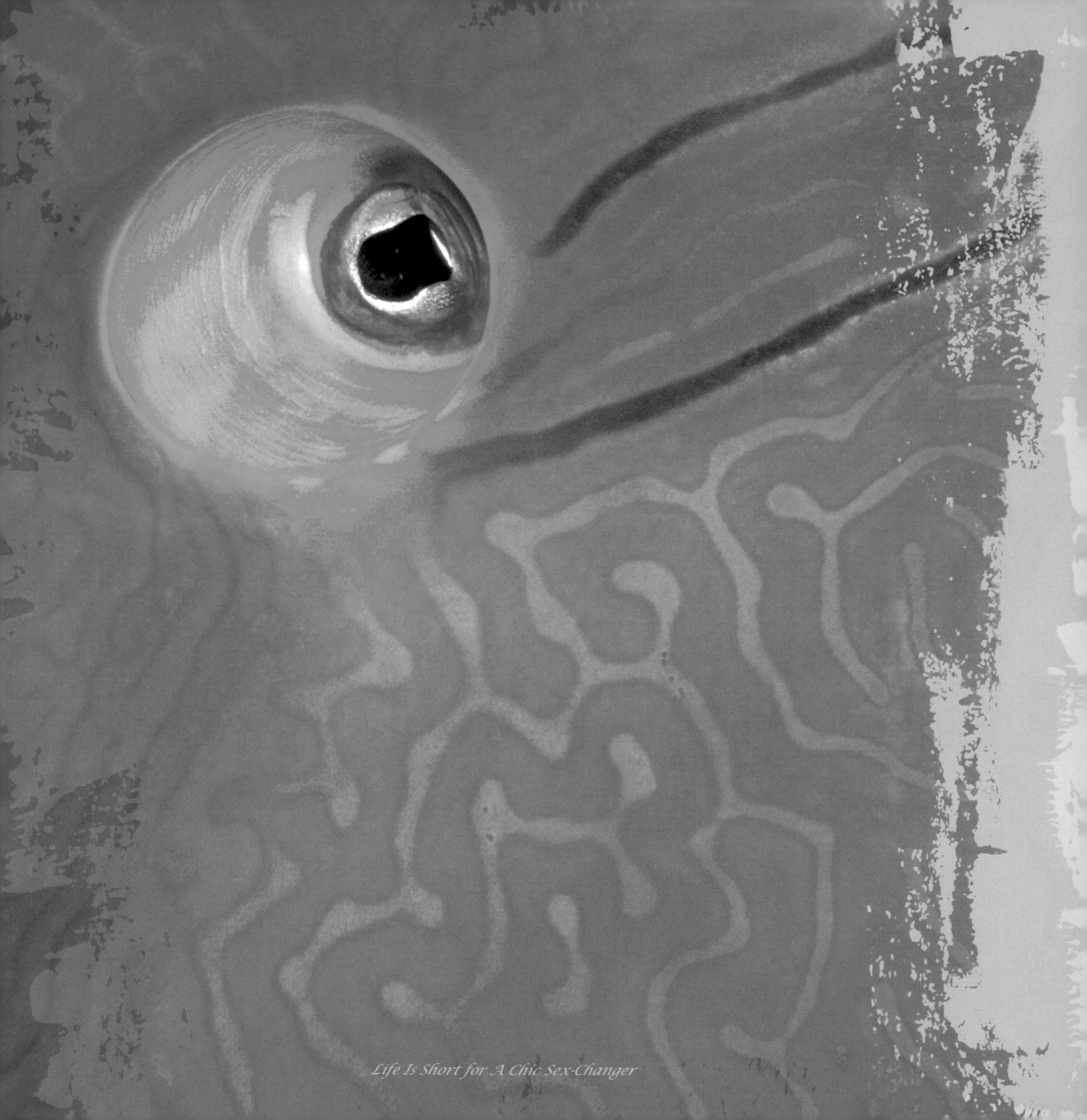

Life Is Short for A Chic Sex-Changer

Life Is Short for A Chic Sex-Changer

The Napoleon Wrasse is a massive and spectacular reef fish threatened by gourmet tastes and the size of dinner plates.

Capable of reaching two metres, it is one of the biggest of all fishes found on coral reefs, and can live more than a quarter of a century. It's also one of the most exclusive of exclusive, expensive gourmet delicacies, and that status is incompatible with the mighty Napoleon's own life-style.

This spectacular animal becomes sexually mature and ready to join other adults to create the next generation at about five years old. Yet it is just around this tender age that a young Napoleon also fits perfectly on a plate. These juveniles are being fished out and dished up at a rate that outpaces the species' ability to replenish itself.

Life Is Short for A Chic Sex-Changer

Even if juveniles do survive to become adults, there remains another challenge to completing their life cycle. This species is one of many reef fishes that employ the reproductive strategy of sex change: adult male Napoleons start out as adult females who then change their sex. No one knows what controls the timing of the change, or why some females change and others never do, but the fish need to live long enough to make the change since it takes both males and females to tango.

Unchecked consumption of juvenile Napoleons could mean an ignominious end for a grand animal that is also known as the Humphead or Maori wrasse. Napoleon is found on the coral reefs of the Indo-Pacific and historically graced the plates of the influential in nations of that region. Considered a "stately" or "royal" fish, it was highly valued in many cultures and used only for special occasions.

But gradually pressures on the fish increased, as divers began to spear it for any and every occasion to sell at any opportunity. The Napoleon also achieved wider fame as luxury seafood; today, it is the highest priced fish in the international trade that brings live coral reef fish into restaurants across Southeast Asia, particularly in Cantonese communities. Captured alive for transport, this fish is worth its weight in silver, literally.

The new and growing demand is from a very large and increasingly wealthy Chinese population. Consumers are not aware that numbers are declining because they still see this fish in their restaurants. In 1996 Napoleons were Vulnerable on the IUCN Red List and by 2004 they were considered Endangered. As Napoleons become rarer their price goes up, and that signals a bleak future for this species. Why? Because our own species invariably really, really wants just about anything that is rare and carries a prestigiously astronomical price tag, never mind the consequences.

Save Napoleon?

In a perfect world, the Napoleon Wrasse (Cheilinus undulatus) *would have a huge worldwide fan club.*

An adult Napoleon is big – about the size of a black bear – and beautiful, colored with electric blues and greens. Its intricately patterned face inspired one of its many aliases, the Maori Wrasse. The fish reaches sexual maturity around age five, not unlike some primates, and can live to be more than 30 years old. Like male tigers, mature Napoleons are loners. And like dolphins, their faces are expressive – Napoleons seem to have personalities.

In other words, they're a lot like the charismatic mammals we love, and that could be a key to their survival.

Grown adults tend to inhabit the same stretch of reef for long periods; many commercial dive sites today have their "resident" Napoleon Wrasse that thrills divers who meet it. Live and wild, where there is a diving industry and ecotourism, this fish can be worth far more than dead ones.

Life Is Short for A Chic Sex-Changer

Listed on the Convention on International Trade in Endangered Species' (CITES) Appendix II in 2004, there is finally a chance that international trade in this species will come under control. Major exporting and importing countries are trying to actively enforce the listing, which has resulted in lower numbers of fish on the market and, hopefully, could give wild populations a chance to recover.

A Side of Cyanide

Techniques used to relieve oceans of life are seemingly limited only by human imagination. But the use of the lethal poison cyanide, which also has been used for human capital punishment, is beyond the pale.

Like many other fishes, Napoleon Wrasse are traditionally taken by hook and line, hand spears, or by trap, depending on the size of the fish. But another more deadly trend has emerged to catch smaller Napoleons alive for the live reef fish trade. Fishermen spray a solution of cyanide pellets dissolved in seawater along a coral reef to knock out the target Napoleons; the fish are then removed to fresh seawater to recover and be eaten another day.

Meanwhile, back at the reef, the lingering cyanide kills corals and countless other unwanted fish and invertebrates exposed to it but not removed.

Turtles Turned Turtle

Turtles Turned Turtle

Turned turtle: an expression as old as time, as old as turtles, perhaps. An unspoken meaning of something turned upside down, of something whose destiny has been reversed, and sadly applicable to its very namesake today.

All seven species of marine turtle on the planet are faced with some form of local extinction, a far cry from their past dominance of the planet's seas.

Sea turtles have been around, relatively unchanged, since the late Triassic Period. Generation after generation of these ancient mariners has navigated invisible routes to feeding and nesting grounds since the age of dinosaurs.

Gliding silently beneath ocean waves, the sea turtle draws on and feels unseen magnetic forces to find its way back home every time. Evolution has left them perfect for their mission: long paddle-like fore-flippers for in-water flight, soft velvety scoops behind to carefully craft baby turtle crèches, and solid backs to carry the world's problems.

Sea turtles spend most of their long lives foraging the ocean's pastures, from seagrass beds to coral reefs and open ocean gyres. Adults migrate hundreds of kilometres to mate in the places where they were hatched decades earlier, and when they arrive females deposit the eggs housing the next generation on familiar dark sandy beaches. All then return to ancestral feeding grounds to replenish and recuperate. They survived the gauntlet of another migration.

And a gauntlet it is. Today, to a sea turtle, the ocean is as much an obstacle course as it is home.

Turtles Turned Turtle

Hungry turtles can be fatally fooled into eating the discarded plastic bags that look like jellyfish in the open ocean. Other garbage is even deadlier. Across oceans drift endless miles of 'ghost fishing nets', discarded overboard when old, or lost to violent storms. These unmanned nets continue to 'fish', endlessly trawling the oceans, entrapping turtle after turtle, victims of innumerable ghost nets drifting unseen in the great blue.

Predatory gill nets and trawl nets all take their deathly toll. Seemingly endless lines of fishing hooks dangle down, ensnaring and entangling. As fishing pressure destroys turtle feeding grounds, turtles themselves are taken inadvertently as bycatch.

Humans who step in to help can often have exactly the opposite effect. Efforts to reduce bycatch have sometimes been at the expense of other species. When fishermen developed a hook design which reduced the number of turtles taken in longline fisheries they found that certain species of sharks were more likely to be taken instead. As if sharks didn't have their own set of problems.

Turtles Turned Turtle

Sea turtles that defy the odds reach the sandy beaches upon which they were born. A female patiently waits offshore for the precise moment she senses all is well to emerge and deposit her eggs. But too often now the right time will never come, as development gobbles up beaches or lights brighten the sands confusing mothers and hatchlings alike.

Turtles have a tough enough time surviving all that nature throws in their path; humans almost unbearably compound the challenge. But the course can be reversed, and already has been in some cases. Some tenacious turtle families have come back from the brink of extinction, though the battle is far from won. Long-term conservation that understands people's needs along with those of sea turtles is the solution; those ancient mariners magically gracing our oceans are the reward.

Climate Change, a Threat to Gender Balance

Gender in marine turtles is temperature dependent, no Y and X chromosomes, no biological game of chance, just temperature. Under natural conditions a turtle laying its eggs in cooler conditions results in more males, and warmer temperatures produce more females. Fortunately nature throws in its fair share of chance, and this has obviously worked well through the last few hundred million years...

But what if climate change gets in the way? What if temperatures get so high that only females are produced no matter where or when they are laid? What if sea level rise means beaches are too shallow for a turtle to nest successfully? Climate change is a reality for turtles; the question is whether their ability to adapt can keep pace.

Turtles Turned Turtle

Conservation Information for Featured Species

The IUCN Red List of Threatened Species

The IUCN Red List of Threatened Species™ is the world's most comprehensive information source on the global conservation status of plant and animal species. IUCN and its partners determine a species' risk of extinction using objective criteria and make the results available to the public through the IUCN Red List online database. Species listed as Critically Endangered, Endangered or Vulnerable are collectively described as 'threatened'. Species that are not threatened fall into either Least Concern or Near Threatened; the other categories are Not Evaluated, Data Deficient, Extinct in the Wild, and Extinct.

But the IUCN Red List is not just a register of names and associated threat categories, it also contains information on what threatens the species, their ecological requirements, where they live, and what conservation actions could be taken to reduce or prevent extinctions.

Among the marine species, IUCN has assessed all known marine mammals, sea turtles, reef-building corals, groupers, and sharks and their relatives, and has joined forces with Conservation International and Old Dominion University to determine the conservation status of all fishes and marine reptiles, seagrasses, mangroves, and certain invertebrates (such as abalone and sea cucumbers) under an initiative called the Global Marine Species Assessment.

The IUCN Red List can be accessed from www.iucnredlist.org.

CITES

The Convention on International Trade in Endangered Species of Wild Fauna and Flora (CITES) is a legally binding international agreement to which 173 governments adhere to ensure that international trade in specimens of wild animals and plants does not threaten their survival.

Species requiring protection from international trade are listed in one of three Appendices: Appendix I contains threatened species for which trade is only permitted in exceptional circumstances; Appendix II is for species where trade must be carefully controlled; and species in Appendix III are protected in at least one country that has asked other CITES Parties for assistance in controlling the trade. www.cites.org

Common Name (English) (in order by chapter)	**Latin Name**	**IUCN Red List Status** (year assessed)	**CITES Appendix**
Southern Elephant Seal	*Mirounga leonina*	Least Concern (2008)	
South African Abalone	*Haliotis midae*	Not Evaluated (Two other abalone species are Endangered and Critical)	III
Nassau Grouper	*Epinephelus striatus*	Endangered (2008)	
Oreo Dory	*Allocyttus niger*	Not Evaluated	
Orange Roughy	*Hoplostethus atlanticus*	Not Evaluated	
Porbeagle Shark	*Lamna nasus*	Vulnerable (2006)	
Spectacled Eider Duck	*Somateria fischeri*	Least Concern (2004)	
Shallow-water reef-building corals	845 species of reef-building corals assessed	One third of assessed reef-building corals are threatened (2008)	II*
Southern Bluefin Tuna	*Thunnus maccoyii*	Critically Endangered (1996 - needs updating)	
Northern Bluefin Tuna (Eastern Atlantic stock)	*Thunnus thynnus*	Endangered (1996 - needs updating)	
Northern Bluefin Tuna (Western Atlantic stock)	*Thunnus thynnus*	Critically Endangered (1996 - needs updating)	
Skipjack Tuna	*Katsuwonus pelamis*	Least Concern (1996 - needs updating)	
Yellowfin Tuna	*Thunnus albacares*	Not Evaluated	
Bigeye Tuna	*Thunnus obesus*	Vulnerable (1996 - needs updating)	
Albacore Tuna	*Thunnus alalunga*	Data Deficient (1996 - needs updating)	
Caribbean Spiny Lobster	*Panulirus argus*	Not Evaluated	
Western Australian Rock Lobster	*Panulirus cygnus*	Not Evaluated	
Ornate Rock (or Red) Lobster	*Panulirus ornatus*	Not Evaluated	
Red Rock Lobster of Mexico	*Panulirus interuptus*	Not Evaluated	
Napoleon (or Humphead or Maori) Wrasse	*Cheilinus undulatus*	Endangered (2004)	II
Flatback Turtle	*Natator depressus*	Data Deficient (1996 - needs updating)	I
Green Turtle	*Chelonia mydas*	Endangered (2004)	I
Hawksbill Turtle	*Eretmochelys imbricata*	Critically Endangered (2008)	I
Leatherback Turtle	*Dermochelys coriacea*	Critically Endangered (2000)	
Loggerhead Turtle	*Caretta caretta*	Endangered (1996 - needs updating)	I
Olive Ridley Turtle	*Lepidochelys olivacea*	Vulnerable (2008)	I
Kemp's Ridley Turtle	*Lepidochelys kempii*	Critically Endangered (1996 - needs updating)	I

*All stony corals, *Scleractinia* species

Resources for More Information and What You Can Do

What You Can Do – Sustainable Seafood Consumption

- *Monterey Bay Aquarium's Seafood Watch: http://www.montereybayaquarium.org/cr/seafoodwatch.asp*
- *World Wildlife Foundation (WWF)'s Sustainable Seafood Consumer Guides: panda.org/about_wwf/what_we_do/marine/our_solutions/sustainable_fishing/sustainable_seafood/seafood_guides/index.cfm*

Relevant IUCN Websites

- *Homepage: www.iucn.org; Species Programme: www.iucn.org/species; Marine Programme: www.iucn.org/marine*
- *The IUCN Red List of Threatened Species™: www.iucnredlist.org*
- *Marine Conservation Sub-Committee: www.iucn.org/ssc/marinespecies*
- *IUCN World Commission on Protected Areas' global web portal for Marine Protected Areas: www.protectplanetocean.org*

Elephant Seals

- *Claudio Campagna: ccvz_puertomadryn@speedy.com.ar*
- *For more information: www.modelo-mar.org*

Abalone

- *Markus Bürgener: burgener@sanbi.org*
- *For more information: www.traffic.org/traffic-bulletin/traffic_pub_bulletin_20_2.pdf*
- *TRAFFIC newsletter featuring abalone trade issues: www.traffic.org/traffic-bulletin/traffic_pub_bulletin_21_1.pdf*
- *Report on issues relating to application of Appendix III of CITES to marine species: www.traffic.org/general-reports/traffic_pub_gen6.pdf*
- *Presentation on abalone trade: www.illegal-fishing.info/uploads/Burgener.pdf*

Grouper Fish Aggregations

- *Yvonne Sadovy de Mitcheson: yjsadovy@hku.hk, scrfa@hku.hk*
- *For more information: www.scrfa.org, www.hku.hk/ecology/GroupersWrasses/iucnsg/index.html*

Bycatch

- *Sarah Foster: s.foster@fisheries.ubc.ca*
- *For more information: www.projectseahorse.org, www.panda.org/about_wwf/what_we_do/species/problems/fisheries_bycatch/index.cfm*

Sharks

- *Sarah Fowler: sarah@naturebureau.co.uk; Sonja Fordham: sfordham@oceanconservancy.org; Claudine Gibson: c.gibson@chesterzoo.org*
- *For more information: www.sharkalliance.org, www.oceanconservancy.org, www.iucnssg.org*

Spectacled Eider Ducks

- *Dorothy Childers: Dorothy@akmarine.org*
- *For more information: alaska.fws.gov/media/SpecEider.htm*
- *Grebmeier, Jacqueline, et al. 2006.* A major ecosystem shift in the northern Bering Sea. *Science, vol. 311, Mar. 10, 2006. Available at: www.sciencemag.org*
- *Krupnik, Igor, ed. 2002.* The World is Faster Now. *Article Research Consortium of the U.S., pp156-197. Available at: www.arcus.org/Publications/EIFN/index.html*

Tropical Coral Reefs

- *David Obura: dobura@africaonline.co.ke*
- *For more information: www.coralreef.noaa.gov/, www.icriforum.org, www.creefs.org/, www.icran.org/index.html, www.coral.org/*

Tunas

- *Eric Gilman: eric.gilman@iucn.org*
- *Majkowski, J. 2007.* Global Fishery Resources of Tuna and Tuna-like Species. *FAO Fisheries Technical Paper 483. Rome: Food and Agriculture Organization of the United Nations. Available at: www.fao.org/docrep/010/a1291e/a1291e00.htm*

Lobsters

- *Mark Butler: mbutler@odu.edu*
- *For more information: www.odu.edu/~mbutler/, www.odu.edu/~mbutlerv/newsletter/index.html*
- *Lobsters: Biology, Management, Aquaculture and Fisheries. 2006. B. Phillips (ed). Blackwell Press, London. ISBN: 9781405126571. Available at: www.blackwellpublishing.com/book.asp?ref=9781405126571*

Deep Sea Corals and Seamounts

- *Dr. Alex Rogers: alex.rogers@ioz.ac.uk*
- *For more information: www.lophelia.org, www.eu-hermes.net/, censeam.niwa.co.nz/, seamounts.sdsc.edu, http://oceanexplorer.noaa.gov/explorations/explorations.html*
- *Rogers AD, Clark MR, Hall-Spencer JM, Gjerde KM (2008)* The Science behind the Guidelines: A Scientific Guide to the FAO Draft International Guidelines (December 2007) for the Management of Deep-Sea Fisheries in the High Seas and Examples of How the Guidelines May Be Practically Implemented. *IUCN, Switzerland, 2008. 39pp. Available at: iucn.org*
- *Freiwald A, Fosså JH, Grehan A, Koslow JA, Roberts JM (2004)* Cold-water coral reefs: out of sight-no longer out of mind. *UNEP-WCMC. Cambridge, UK. 84 pp. Available at: www.unep-wcmc.org/press/cold_water_coral_reefs/report.htm*

Napoleon Wrasse

- *Yvonne Sadovy de Mitcheson: yjsadovy@hku.hk*
- *For more information: www.humpheadwrasse.info, www.hku.hk/ecology/GroupersWrasses/iucnsg/index.html*

Marine Turtles

- *Nicolas Pilcher: npilcher@mrf-asia.org*
- *For more information: www.seaturtle.org, www.ghostnets.com.au, or www.mrf-asia.org*

Acknowledgements

This book would not have been possible without the generous and enthusiastic support of many colleagues and peers in the conservation world.

Much credit is due to all the members of the IUCN Species Survival Commission Marine Conservation Sub-Committee (MCSC), from which the concept for this book emerged. Special thanks goes to Holly Dublin, at the helm of the SSC, for her vision to create and support the MCSC. Jane Smart, who leads the Species Programme, is also specially thanked for making valuable time and resources available for this project.

For their outstanding artistic contributions, we thank and acknowledge Eugenia Zavattieri for lay-out and design, and Alison Power for writing stories based on the technical submissions of our expert contributors. We also would like to thank all the story contributors and photographers, and our colleagues who helped in many other ways: Michael Nolan, Carol Foster, Richard Foster, Brice Semmens, Jason Isley, Claire Nouvian, Bill Herrnkind, Imène Meliane, James Oliver, Cindy Craker and Deborah Murith.

No project can run on inspiration and creativity alone, funding is always a necessity, and we are grateful to the donors for their generous financial and in-kind support: the Lighthouse Foundation, the European Bureau for Conservation and Development, the University of Hong Kong, the Society for the Conservation of Reef Fish Aggregations, the Marine Research Foundation, the Wildlife Conservation Society, IUCN Species Programme, IUCN Species Survival Commission, and IUCN Global Marine Programme.

SSC Marine Conservation Sub-Committee

Claudio Campagna, *Wildlife Conservation Society Sea and Sky Project and National Research Council of Argentina, Argentina (MCSC Co-Chair)*
Markus Bürgener, *TRAFFIC, South Africa*
Dan Laffoley, *Natural England, UK*
Christoph Imboden, *D. Swarovski & Co, Switzerland*
Stefan Nehering, *Aqua et Terra umweltplanung, Germany*
Nicolas Pilcher, *Marine Research Foundation, Malaysia*
Alex David Rogers, *Institute of Zoology, Zoological Society London, UK*
Yvonne Sadovy de Mitcheson, *University of Hong Kong and Society for the Conservation of Reef Fish Aggregations, Hong Kong (MCSC Co-Chair)*
Despina Symons, *European Bureau for Conservation and Development, Belgium*
Marcelo Vasconcellos, *UN Food and Agriculture Organization, Italy*
Amanda Vincent, *Project Seahorse, Fishery Centre, University of British Columbia, Canada*

SSC Marine Conservation Sub-Committee Observers (IUCN Staff)

***Jane Smart**, Head, Species Programme*
***Andrew Hurd**, Deputy Head, Global Marine Programme*
***Kent Carpenter**, Global Marine Species Assessment Coordinator, Species Programme*
***Julie Griffin**, SSC Support Officer, Species Programme*

Contributors

***Markus Bürgener** (Abalone)*
Senior Programme Officer, East and Southern Office,
TRAFFIC (the wildlife trade monitoring organization)

***Mark Butler** (Lobsters)*
Professor of Biological Sciences, Old Dominion University (USA)

***Claudio Campagna** (Elephant seals and Editorial Team)*
Director, Wildlife Conservation Society Sea & Sky Project; Researcher, National Research Council of Argentina;
Deputy Chair, IUCN SSC Pinniped Specialist Group

***Dorothy Childers** (Spectacled Eider Duck)*
Fisheries Programme Director, Alaska Marine Conservation Council

***Sarah Foster** (Bycatch)*
PhD Candidate, Project Seahorse

***Sarah Fowler** (Sharks)*
Managing Director, NatureBureau International; Co-Chair, IUCN Shark Specialist Group

***Sonja Fordham** (Sharks)*
Director, Ocean Conservancy Shark Conservation Program; Deputy Chair, IUCN Shark Specialist Group

***Claudine Gibson** (Sharks)*
Programme Officer, IUCN Freshwater Fish Specialist Group; Former Programme Officer,
IUCN Shark Specialist Group

***Eric Gilman** (Tunas)*
Marine Science Advisor, IUCN Global Marine Programme

***Julie Griffin** (Project Coordination and Editorial Team)*
SSC Support Officer, IUCN Species Programme

Andrew Hurd *(Editorial Team)*
Deputy Head, IUCN Global Marine Programme

David Obura *(Tropical coral reefs)*
East Africa Coordinator, Coastal Oceans Research and Development – Indian Ocean (CORDIO);
Chair, IUCN Global Marine Program's Climate Change and Coral Reefs Working Group

Nicolas Pilcher *(Marine turtles and Editorial Team)*
Founder and Executive Director, Marine Research Foundation; Co-Chair, IUCN Marine Turtle Specialist Group

Alison Power *(Writing)*
Freelance popular science writer

Alex David Rogers *(Deep sea corals)*
Senior Research Fellow, Institute of Zoology, Zoological Society of London

Yvonne Sadovy de Mitcheson *(Grouper fish aggregations, Napoleon Wrasse and Editorial Team)*
Professor, University of Hong Kong; Co-founder Society for the Conservation of Reef Fish Aggregations;
Chair, IUCN Groupers and Wrasses Specialist Group

Eugenia Zavattieri *(Artwork and design)*
Artist and artisan

L'IV Com Sàrl *(Production)*
Graphic design firm

Photo Credits by Chapter

Hanging at Sea

Elephant Seals

***Jim Large**: p.8, p.10 / **Guillermo Harris**: p.11 / **Claudio Campagna**: pp.12-13, p.15 top right and bottom right*
***Victoria Zavattieri**: p.14, pp.16-17 / **Marcela Uhart**: pp.14-15 top center*

Drugs, Money, Madness – and a Mollusc

Abalone

***Rob Tarr**: p.18, p.22 main / **Markus Bürgener**: p.20, p.22 right / **James Middleton**: p.18-19 center, pp.24-25*

Fatal Attraction for Fish

Nassau Grouper

***Seapics**: p.27, p.32, p.34-35 / **Doug Perrine**: p.28, pp.30-31 centre, top right, and bottom right /*
***Pat Colin**: pp.30 left*

Waste and Horror at Sea

Bycatch

***Sarah Foster**: p.36, p.38 pp.40-41, p.43, p.44 / **Stanley Shea**: p.42*

Man Eats Shark, World Yawns

Porbeagle Shark

***SeaPics**: p.46, p.48 / **Thierry Minet**: p.50 left side- top, middle, and bottom /*
***Justin Ebert**: p.50 right side- top left and bottom, pp.54-55 / **Janine Mapurunga**: p.50 right side- top right*
***Eran Kampf**: p.51 / **Jane Dermer, Dhimurru Rangers**: p.52*

Ducks Huddle Up to Create Secret Arctic Hideaway

Spectacled Eider Duck

***Drs. Flinson** (Flickr user-name): p.56 / **Bill Larned**, US Fish & Wildlife Service: p.58 top /*
***http://www.cls.fr/html/argos/documents/newsletter/nslan51/spectacled_eider_en.html**: p.58 bottom /*
***http://www.absc.usgs.gov/research/Banding/images/Bering_wintering**: p.58 background /*
***U.S. Fish & Wildlife Service**: p.59 / **David La Puma**, david@woodcreeper.com: p.60 overlay /*
***Jane Beacham** (Flickr user-name: mrsminib): p.62 top and bottom /*
***Will** (Flickr username: dayglowill): p.62 middle / **Matthew Cullen** (Flickr user-name: mcullen): p.65*

Searching for Weapons of Mass Destruction? Visit the Nearest Coral Reef
Shallow corals
***Jerker Tamelander**: p.67, p.69, p.72 right, pp.74–75 / **Jurgen Freund**: p.70–71 all photos /*
***Adriana Antelo**: p.72 left*

Reaching New Heights or Sinking to New Depths?
Tunas
***SeaPics**: p.76, p.79, p.82, p.83, p.84 both, p.86 overlay / **Ramon Benedet**: pp.80–81 both /*
***Aziz T. Saltik**: p.86 background*

More Than an Entrée
Lobsters
***SeaPics**: p.88, pp.90–91, p.92, p.94 bottom right / **Lee Weller**: p.94 middle left / **Chris Seher**: p.94 bottom left /*
***Gaetan Lee**: p.94 top right / **Gwen K. Noda** (Flickr user-name: f_iodinea): p.97*

The Last Frontier, Destined for Devastation?
Deep-water corals and sea-mounts
***Les Watling**: p.99, p.100 / **Alex David Rogers**: p.102, pp.104–105*

Life Is Short for A Chic Sex-Changer
Napoleon Wrasse
***SeaPics**: p.106 / **Oriolus** (Flickr user-name): p.108 background / **Leonard Low**: p.108 overlay /*
***Valerie Ho**: p.110 / **Liu Min**: p.111 / **Amaneshi** (Flickr user-name): pp.112–113 / **Ariffin Aris**: p.114*

Turtles Turned Turtle
Sea Turtles
***Nicolas Pilcher**: p.116, p.118, pp.124–125, p.126 bottom, pp.127 / **Matt Oldfield**, scubazoo.com: p.120 /*
***Gary Luchi**, ghostnets.com.au: p.123 top / **ghostnets.com.au**: p.123 bottom left /*
***Vance Wallin**, Queensland Turtle Conservation: p.123 bottom right / **Bivash Pandav**: p.126 top*

Note: Photographers credited with their Flickr user-name can be found at www.flickr.com